SCARS

ANASTASIOS
TSONIS

Some very short stories
Or maybe (random) poems

DOS MADRES
2023

DOS MADRES PRESS INC.
P.O. Box 294, Loveland, Ohio 45140
www.dosmadres.com editor@dosmadres.com

Dos Madres is dedicated to the belief that the small press is essential to the vitality of contemporary literature as a carrier of the new voice, as well as the older, sometimes forgotten voices of the past. And in an ever more virtual world, to the creation of fine books pleasing to the eye and hand.

Dos Madres is named in honor of Vera Murphy and Libbie Hughes, the "Dos Madres" whose contributions have made this press possible.

Dos Madres Press, Inc. is an Ohio Not For Profit Corporation and a 501 (c) (3) qualified public charity. Contributions are tax deductible.

Executive Editor: Robert J. Murphy

Illustration & Book Design: Elizabeth H. Murphy
www.illusionstudios.net

Typeset in Adobe Garamond Pro
ISBN 978-1-953252-96-8
Library of Congress Control Number: 2023950501

First Edition

Published by Dos Madres Press, Inc.

ACKNOWLEDGEMENTS

I like to thank my daughter Michelle
for being there and always my inspiration,
and Michael Branks for his encouragement.

to

Γ

Table of Contents

Some scars are good
Some not
But they are all reminders…

The soul carriers

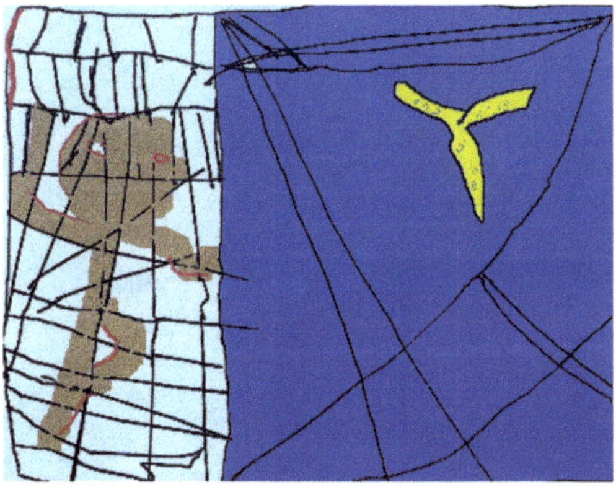

Hey brother
Would you know?
My soul is carried away
    By something like a butterfly
She is taking me around
Like a taxi
Not like a New York yellow taxi
Or something like that
But as a carrier
Which takes me around
Looking over
What I left behind

Ode to an identical twin

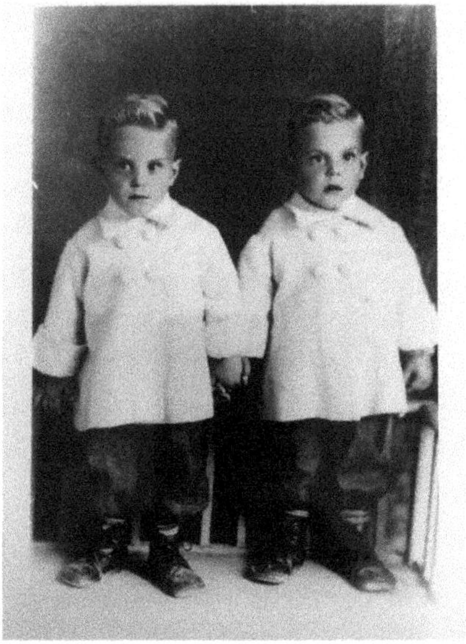

The softer and the harder
The sensitive and the indifferent
The emotional and the reasonable
The calm and the upset
The poet and the writer
The weaker and the stronger
The unlucky and the lucky
I cheated you
I took more
And you were left with less
You were the weaker
I was the stronger
I cheated you
But you never complained
Because you loved me

Even a small uphill creates a noticeable wind
How then can I not notice you?
Fading face
Eyes drifting into the darkness of future
Come find me you say
I will wait for you, you say
You took the moon away
And now it's always dark at night
A rainbow missing one color
You touch me and I feel I exist
I want you back
Just for a second
A second
Too much to ask?
Parallel love
Infinite the distance
But for one moment
I wish
You cross to my universe
And touch me
Just for the smallest unit of time
Whatever that is
I dance for you
A zeibekiko
In the space-time floor
You in some other universe
Suspended in time
Looking at me through our photo together
Laughing at my style
Because I want to be funny
To make you smile
In that parallel line

Which will never meet my line
But really who cares?
We are very close parallels
We live for ever
Passing from universe to universe
And are together
In every dream

Time

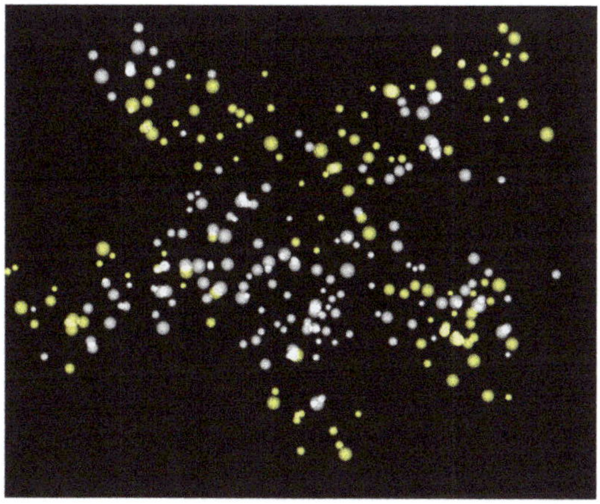

Not a star
Not a black hole
Not a wormhole
Not strings
Not atoms
Not planets
Not comets
You are space
And time
Some say
That time is a dimension
Different from those of space
Which does its own dance
On the floor of the Universe
Time's dance is slow
It is lonely
But others say
That time is a dimension

Entangled with those of space
Two entwined colors
Dancing together
On the floor of the Universe
In the continuum of infinity
A complex dance
Yet others say
That an even more mysterious dance
Permeates the Universe
A dance so complicated
That boggles the mind
A dance which when understood
Will open the doors of the Universe

The game of the name

Tell me a story from your mouth

Today I have a good one for you

I am all ears

Scientists contend that there are more than three spatial dimensions
In fact, they believe that there are at least six more spatial dimensions
But during the Big Bang (which as you know ignited our Universe) they did not make it

What do you mean they did not make it?

Well, I mean that they did not manifest themselves
As did the three dimensions we live in
They simply curled in extremely small scales
And created some kind of a 'sub-space' universe

This is indeed very different story

In this sub-universe
Like in our universe
Life developed but in a tiny little planet
Extraordinarily, the people on this planet were very similar to us
They actually looked like us
The only difference was that their social evolution trajectory
Took 'a strange turn'

A strange turn?

The way their society evolved was as follows
There was a place called 'The City' where most people lived
But there was another place called 'The Club'
Where only selected ones could live
What is funny is the procedure they adopted
To allow people from 'The City' to enter 'The Club'
The procedure was called
'The Game of the Name'

You mean
'The Name of the Game'?

No, I meant what I said, 'The Game of the Name'
You see, the challenge was to present yourself
In front of a committee and tell them your name
The name must have been so amusing as to make the
committee laugh
Then you were allowed to enter 'The Club'
You were given privileges, and a comfortable living
It was everybody's dream

So, all your success and life were based on being creative
in finding a funny name?

Why not?

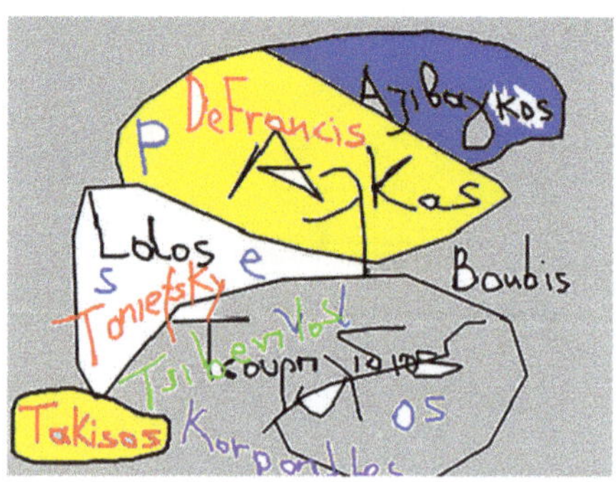

Dream

She walks on gravel
Each step sounds different
She recalls the round room surrounded with doors painted
in different shades of brown
In her dream the room begins to spin
The doors one by one merge together
When the room stops spinning there is just one door
She hesitates but eventually turns the knob slowly
The door opens and reveals a long corridor
A corridor decked with paintings
She walks with caution
She realizes that all the paintings are identical
A girl crouching, looking ahead with emotionless eyes
Looking into a void
To a solid house, which also seems haunted by the void
A house unable to recall moments lived in it

Love, sex, fights, cooking, parties, music, all never existed for it
Like nothing ever happened
The sun is hiding
No trees are visible in the background
Not a cloud in the sky

Abstract poem

In this room there is room only for your body
Imagine
The moon in your ring
Two feathers of an eagle
Close your eyes
It's just another storm
Sleepy lights
Looking forward
Two colors in the rainbow are missing
Can the color of your eyes discriminate?
Night from day?
Day from night?
Yellow from orange?
Orange from yellow?
Long road

From north to south
Lost in the way to love you
From east to west
I can't find a way
To make you love me
Except in my dreams
No compromise
If you need me
Ask the wind what is the street
Without return
Do something
To miss my ride
Throw your body in a glass
Let me drink you one drop at a time
Count the kilometers one by one
Irreparable damages have no returns
Only regrets
How can we imprison thousands
Of 'I love you'?
By becoming statues?
By locking them in a drawer?
There is no absolute darkness
Traces of light are everywhere
Traces
In the light of the stars
Tears, drops in the rain
Maybe you know
But you don't speak
You cannot hide
You can destroy
But you cannot erase dreams
Returns

Disasters
Late sunrise
Sudden sunset
Sad full moon
The random dance of night flies
The problem with parallels is that they never get together
A lie
Loneliness, give me a minute
To go to infinity
Where memories escaped
Where no lies lie
Tell me something new
I like it
I mean the longing
Not the lie
C
    l
    o s
        e
    the
            d
            o
            o
                r

Scars

Can you please come to be with me tonight?
It's late…
Please come and I will cook you a lobster tail
The scar on her spine
The mark of chemotherapy
I feel ashamed that I ever said that it's late
When it's time to go to bed
She asks me if I can be next to her while she falls asleep
I lay next to her and I whisper
You are the only thing in my life
Please do not to leave me
I will never leave you daddy
And then I just sat there by the bed for hours
While she slept
Thinking, why did I ever say it's late…

D^3

Random thoughts

What is square one?
The starting line
And square two?
The first step
The one you are quidded through
As you are for
The second step
And then for
The third step
But there will be a step
That you have to take on your own
Do I always have to take a step forward?
Not always

Sometimes you have to hold your step
And think before you take it
Calculate the possibilities
And if it does not look good
Take a side step
Life is not one dimensional

What is created when intelligence is created?
Do mosquitoes think that when they are smashed to
death they go to heaven?
Or hell?
Do they care?
Then why do we care?
Is it because intelligence created a trap?
In a way a trap on itself?
Come on God, wake me up
Give me a hint
North wind?
Some kind of a hint
Empty bottles
What is a single tiny cloud doing alone in the sky?

A wasp is in my room
Walking around me
Becoming friends
After I killed her colony
A fly circling an upright lamp
Attracted by its blinding light
Not able to avoid the fatal attraction
The light dyne
Plunging it to its death
We are all musical notes

We impress each other with notes
We love in sync
We hate out of sync
We die in between
If we are here by chance, what is chance?
If we are here for a reason, what is the reason?
Is chance then reason's twin?

Dream

Visions of drifting to wastelands
Darkness obscuring all beauty
Crouched on my knees with crows as my company
Crows, ugly crows, noisy unpleasant crows
Cursed to never have a melody coming out of their larynx
And agony spilling out of my face
Accumulating on the ground like dust on the floor
Dust that sticks on your feet and then is carried everywhere

A war of shadows

Gallery room

A black and white house
In a black and white village
Worlds within worlds
A study in nature
Two bodies made out of pure geometric figures
Of triangles and circles
Looking at each other
Possibly conversing
A race track like a bull's head
A rabbit imagining a habitat in the clouds
White spaghetti on a black floor
Two eyes that think they are the same
Five bullets aimed at five abstract boards
Fishermen returning
Black random lines on a white background
White random lines on a black background
Plus to minus
I will come to your dreams
Either as black lines on a white background
Or as its minus
Forming an abstract body
With a trembling hand
Like a small e a r t h q u a k
e

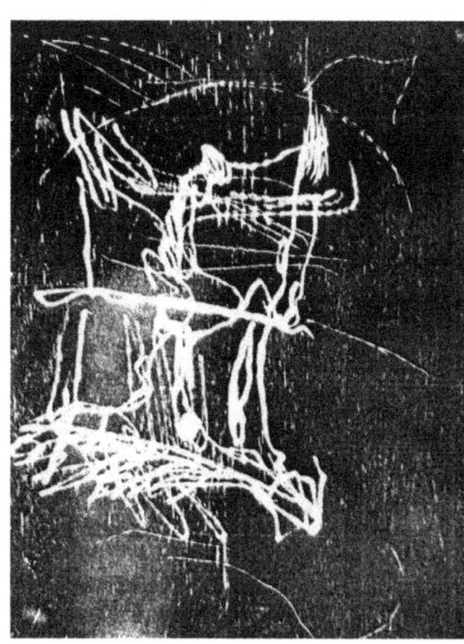

Time

Talking about time
What time is it?
4:30
4:30 what?
4:30 pm
When?
June 6
June 6 when?
2022
And in a year from now it will be 4:30 pm June 6, 2023
And what makes a year?
Earth's motion around the Sun
This motion is very regular and it takes 365 and ¼ days
to complete
It looks like this

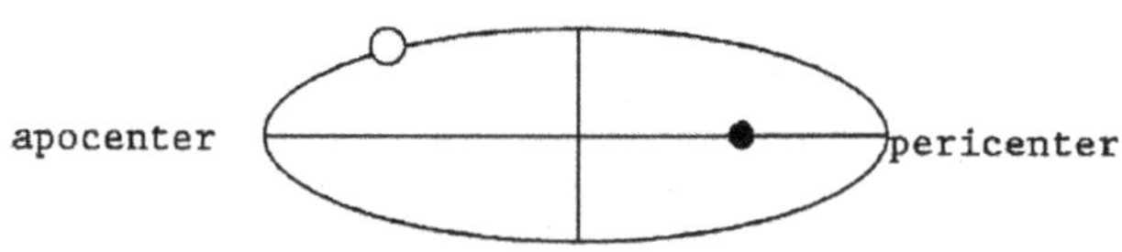

It's periodic, meaning time is exact
But what if this motion was not regular?
What do you mean?
Have you ever heard of the three-body problem?
No
Well, here it is

Imagine the system of Sun, Earth, and the Moon
The Moon is too small compared to either
the Sun or the Earth
As such it does not affect the motion of Earth
around the Sun
But the Moon is subjected to the gravitational pull
of both Earth and Sun
It goes around the Earth which goes around the Sun
What is, in this case, the orbit of the moon around the Sun?
I have no idea
It turns out that depending on the size
of the bodies involved
The trajectory of the three-body system can become chaotic
What do you mean by chaotic?
It means not regular
Not periodic
The trajectory loops around but in a complicated way
Never repeating itself
And never crossing itself
It may come very close to a past position
But never to a point of the past

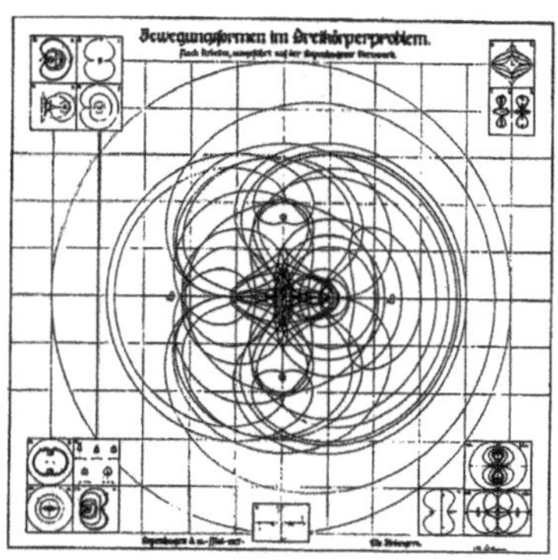

In other words, 4:30 pm June 6 will never happen again
Are you serious?
Do I look that I am not?
Then, in a system like that
Time cannot be measured the way we measure it here
How can we then measure time?
Good question
A butterfly lives 2-3 weeks
We live around 80 years
Does the butterfly think time is unfair to her?
What is the time domain of a butterfly?
Time on base 10?
Or maybe base 2?
Time is relative
Like 'download time'
The computer adjusts it as it downloads documents
Time can stand still
Time flies
(when you have fun)
Time takes too long
(when you wait)
Time can work things out
Time can heal
(so they say)
Time can cheat you
You cannot cheat time
Time can be your friend
Or your enemy
You cannot kill time
Time has its ways with you

Past

Things happened
Good things
Bad things
Accidents
Random things
Right decisions
Wrong decisions
Joy

Sadness
Elation
Disappointment
You name it

Present

Only lasts for an infinitesimal interval of time
Before you know it, present is past

Future

A black void

But all three in the realm of infinity

In their nature

How do you break loneliness?
Drink?
Smoke?
Dig in memories?
But then again why break it?
How do you hide
A tear from your eye?
How do you hide a sad eye?
But then again
Why hide it?
Not your fault
It's your nature
Birds over deserts
Desperately looking for water
But why are they there?
Forgotten?
Lost?
Attached to a past?
Never letting go?
Part of their nature?

Up in the sky
Insight a cloud
There is a fight
Negative and positive
They don't like each other
Unavoidable connection
Loud noise
The cloud cries
And then calm
Only to repeat
But
Why do clouds cry?
Because they have no choice?
Because it's in their nature?
Echoes of the west
Shadows of the east
North and south in constant contrast
Endless motion
Obeying rules
Nature's way

Information on trial

Imagine the following

You are inside a dark room
No windows, no cameras, no recordings of any kind
You write in a piece of paper something that you only know
Then you burn the paper
Mix the ashes
Get out of the room
Go in the middle of Pacific Ocean and spread the ashes

Can the information written on the piece of paper in that room
ever be recovered?

Yes, if you talk when you sleep and your partner hears it

Okay, change the scenario

You drown in the ocean with the ashes

Now imagine two identical containers filled with water
in exactly the same way
Drop the same amount of red ink in both of them
but at different points
What will the final result be?
If we wait long enough the ink will have diffused uniformly
in both containers
The two containers will look identical

How, in this case, can one recover information about the initial
position of the ink dropping?

Another example

Imagine a compartment where a gas
Is confined by some diaphragms in one of the corners
With its molecules moving around randomly

You then open the diaphragms and let the gas move
What will happen?

The gas will expand and soon will occupy the whole box
This is the basis of the second law of thermodynamics
The supreme law of nature
It basically states that the entropy of the system increases
That probability will have the gas occupy the whole volume
And that it will take some lives of the universe for it to
By chance
Find itself back to its original confinement
This

For all practical purposes
Means that the initial information about the location of the gas
is lost
Thus
The second law implies that information can be and is lost

Now, one may argue that this is not absolute
Because there is always a probability
That the gas will converge back to its initial volume
But will the molecules also assume the same position?

Seriously?

But really, what is probability?
How is it calculated?
Probability is based on past history
If you have all past mean temperatures of January
You can calculate how many times
That temperature was above or below a certain value
And that allows you to assign a probability to exceed or be below
That certain temperature

Now what if the probability of something happening
Is longer that the age of the universe?
If probability is based on the past history
It means that in this case probability cannot be defined
Which is a conundrum
Thus, if invalidating the second law will take longer than the age
of the universe
Then you cannot invalidate it

The Universe has other things to worry about…

Random thoughts

Math and love
Or Love and Math

The distance of -3 from zero is 3
The distance of +3 from zero is 3
The absolute value
Represents the distance from zero
Which is always positive
It equalizes minus and positive
The absolute value of -3 is 3
The absolute value of 3 is 3
One stays as is
The opposite changes
Absolute love

Equalizes opposites
Or is it?
Does it mean that absolute love
Will change one
While the other remains as is?
Is this fair?
What is zero in love?

Senses

They say we have five senses
Hearing, eyesight, smell, touch, taste
(they also talk about the sixth sense, but let's not go there now)
Some can see better than others
Some can hear better than others
And so on
But how about
The sense of time
Or that of space
Some are never on time
Some are always on time
Some never get lost
Some can never find their way
(of course, now they can always use Google maps)
Space/time is in our genes
Some manage it
Some don't
So, really there are seven senses
(and maybe an eighth, but let's not go there now)

The three points

In sports
It is lately a procedure to give some kind of interview after
the game
What has become the major point in my life
Is what in one of those interviews a soccer player in Greece
Said about the game
"Well, we did not play as we planned
We made some mistakes
but we won
We made it
And that's what counts
"We keep the three points"
Things come to an end
And usually
The end may have a degree of sadness
What makes the balance in your favor
Is the good elements before the end
What was good and what was worth it
Those that you keep
To tell you that in the end
It was all worth it
The three points
The six keys on the counter

Hahaha

When he goes down the drain
Does the drain know?
When he goes down the stairs
The stairs don't object
When he goes in an airplane
The airplane lets him fly
When he is in the kitchen
He does not necessarily cook
When he is in the garden
He accidentally steps on worms
When he gets in his car
Does the car worry
That this would be its last day?
When he pets his cat
The cat loves it
When he appears aloof
Is he really aloof?
Or in his own world?
When he looks at you
With his piercing eyes
What do you feel?

Random song

Random screens
With your face on them
Rising in my path
Uninhabited pictures
Tears are waste of water
Ruined farm houses
Perfect subjects of modern photography
Why will a bird not want to sing?
Before the door
I pray
Not to God
But to Probability
Not to find
What I fear
Cold bed
Snow on green grass
Snow on olive trees
Motionless leaves
No moon in the sky
If I lose you
Every tear
Will become a storm

A gap in history
Lost in the wrong riverbank of time
Where nothing is important
Before the door
I pray
Not to God
But to Probability
To find
A corridor just for us
An abstract flower
Unlike any other flower
A hummingbird looking directly at me
While hovering in front of me
Moving in straight lines
And in right angles
Fireflies spinning in perfect circles
No storms
Just the edges of clouds
Dancing before disappearing
Three moons in the sky
No gaps in history
Time next to me
Faintly shining

m y p a t
h

Come to me
Like merging clouds
Not like merging galaxies
That takes too long

Why?

Do

Rabbits

Fall

In

Love?

Imagine

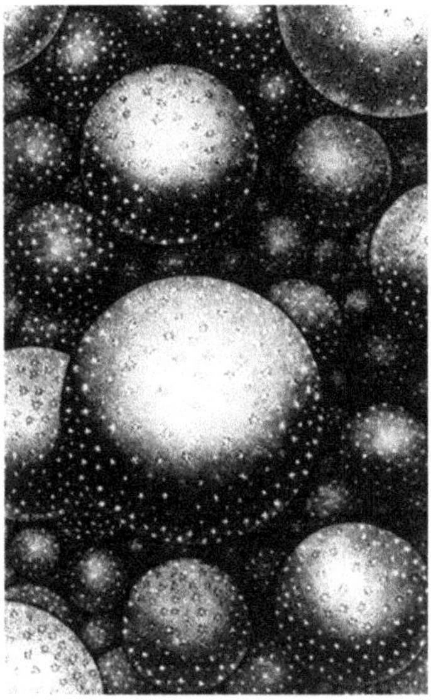

An endless sunshine
An animal touching you softly
A bird walking with you
Two galaxies merging
Infinite universes
Parallel universes
Where all choices exist
Where nothing ever went wrong
A dream
You would never want to forget
A dream
You would not want to wake up from
A dream
You would want to assimilate with
A dream
Where I hold your hand again

A dream
To transform you
A dream
To take you to a parallel universe
Where I will know you
All over again
Where we walk in rain
And maybe in snow
And definitely in sunshine
Let the light clean everything
To correct

errors

Plus, or minus?

Disappearing wolfs
A lonely bird on a power line
A sinking ship
A deserted house
A hidden sunrise
Flowers withering
A lonely man in a street
A woman walking the opposite direction
Increasing their distance
Decreasing their time
An inferno of unwelcomed
Helpless feelings
As is the wind
Which disperses them everywhere
What is a single cotton ball doing alone in the sky?
On
The
Other
Hand

Wolfs are back
A flock of birds are discussing (something) on a power line
A ship sailing under a full moon
A house full of lights
An unknown flower
Coming to life
Under a promising sunrise
A lonely man in a street
A woman walking toward him
Decreasing their distance
Increasing their time
Feelings slowly flaring
And the wind helps
By engulfing them
Is
The
Glass
Half
Empty
Or
Half
Full?

Ode to an identical twin

Probability took you
If A did not happen
B would have not happened
If B would have not happened
C would have been in your favor
But still if
D did not happen
You may still be here
But in your last breath
You had to do D
Because you loved E
Only T could do this

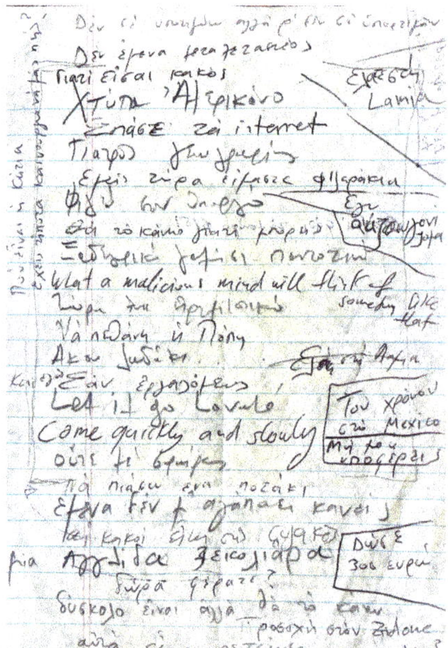
What a malicious mind will think of
Let it go Lovato
Come quickly and slowly
Του χρόνου στο Μεξικό
Δώσε 300 ευρώ

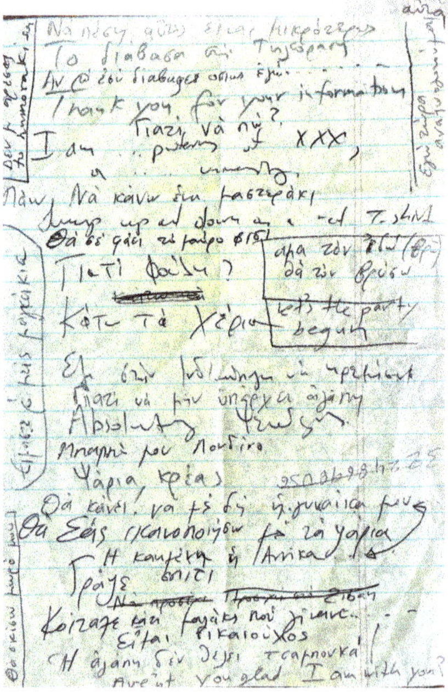
Thank you for your information
Γιατί να πω?
Jump up and down
Γιατί φοβάσαι?
Κάτω τα χέρια
Let's the party begin
Ψάρια, κρέας
Aren't you glad I am with you?

When I will cry
I don't know if I will laugh
I don't know if
I will remember you
He should apologize
He is the younger
But it's your fault
It does not matter
He is the younger
Nobody loves me
Aren't you glad I am with you?
Why are you mean?
I don't underestimate you
But I don't overestimate you either
(I guess mid-way is best)
The bad ones are in jail
I read it on TV
I am thinking of going to England
To do a little Master of Science
Bullying and love don't go together
(Η αγάπη δεν θέλει τσαμπουκά)
Jump up and down in a red T-shirt
Doctor of Geography
And now something relaxing
Forget the internet
We are best friends
You will do it
Because you want and you can
Come quickly and slowly
You brought presents?
It's difficult boss, but I will do it
Attention to Zidane

Come to Indianapolis to relax
Why should not there be love?
We are little macho too
Let's the party began
Thank you for your information
I am now revitalizing
I don't like your little humor
Why Fouli?
Right Katia?
And from now on fish and meat!
Kisses to the minister
Ointment?
What ointment?
Why ointment?
Oh, shut up
Why should I?
Hands down!
I wish I did not make the grade
Oh, come on
Just give me 300 euros
Let it go Lavalle

Dream

In my dream I am surrounded by white
Everywhere I step is white
Everywhere I look is white
And it is cold
The definition of emptiness, I say to myself
And it is so cold…
White… A color of hope and at the same time of agony
A color of purity
But so easily spoiled
When all is white you can see nothing
But blue?
"*Now that's a color*", he would say…
"*First snow always melts*", he would also say…
Dreams…

Untitled

You remind me of somebody
I hope somebody dear
Oh yes, very dear
What happened?
I lost her
No, you didn't
What did you say?
No, you didn't
Repeated the reflection in the mirror
In math
A 'closure'
Refers to having achieved a solution
Which has not loose ends
Which leaves no room for speculations
Or 'snakes in the grass' to question its validity
Mirror, mirror

Is there a closure to suffering?

NIGHTHAWKS

And when all is said and done
The picture will be simple
And clear as ever
Reality will seem like a dream
And dreams will seem like reality
It will not be foggy
It actually
Will be very sharp
And peculiarly inviting
But it will be lonely
Hiding our insecurities
Sometimes I think love is a curse
A joke the devil invented
To laugh at people
While waiting for them to join him
But there is no devil
Only wandering hearts
In empty streets
Birds of paradise
Yellow, green, and red

Untitled

I think of the desert
Don't ask me why
One of my many thoughts
A quiet place
A monotonous place
A remote place
A hot place
Such an exotic place
Such a desperate place
Such a lonely place
Do you know that flowers grow in the desert?
Yes daddy
If they can survive adverse conditions so can we
Yes daddy

Entanglement

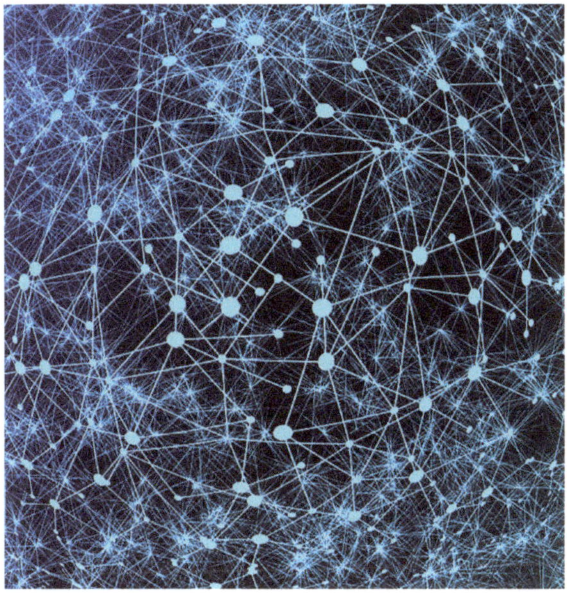

Wonder in your eye
Never a tear
A gaze is lost
Travels across
The cosmos
Infinite thoughts
Chasing you
I, the wanderer
Gird with light
Find you instantly
Because my mind
Dances in tunes
That know no master
In parallel

We walk in parallel
And it all is restricted
For us to joint together again
Parallel love
Leading me
In a road named Errors
In parallel
We walk in parallel
But our minds
Don’t follow our steps
Our minds interweave
Lifting all restrictions
Don’t be afraid
I am with you
Nothing bad will happen to you
Stay in our space
And in our time

Clouds
(Or an alternative to weather reporting)

The clouds are crying
(It's raining)
The clouds are upset
(Lightning)
The clouds are sad
(Gloomy overcast)
The clouds are happy
(Small cumulus dotting the sky)
The clouds are cold
(It's snowing)
The clouds are absent
(Clear skies today)
The clouds are high and thin
(Can you see the halo?)
The clouds are in agreement
(They cover the sky uniformly)
The clouds seem random today
(Tell me about it)
Some look like birds

The leaf

A single leaf
Alone in the tree
At the mercy of the wind
Spinning
In a random geometric pattern
That seems like
Right
Down
Up
Left
But in all directions
Amorphous
Complicated
Drawn by
An amorphous
A complicated
Wind
And looking at this
Sparkling eyes
A bird flying out of them

A melting pot of
Excitement
Relief
Feelings
Memories
Good and bad
That drifts you like
An opera song
Sunrise
The leaf is at rest now
Nature is changing her clothes
Spring knocks at the door

Ode to Bob Dylan

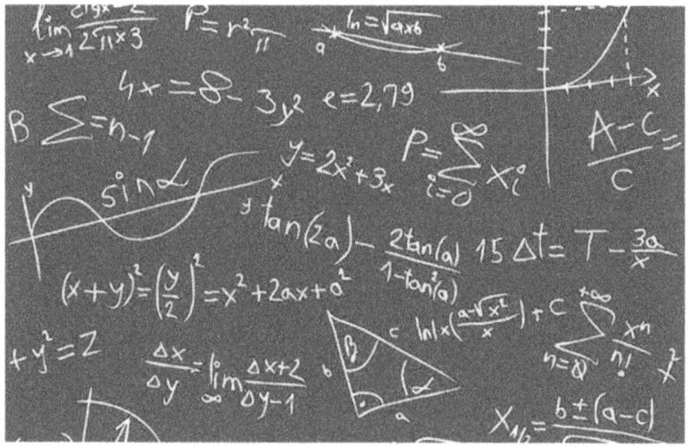

You may be a +
You may be a –
You may call yourself an *x*
Or you may call yourself a *y*
But altogether in an equation
You have to obey some rules

You may be real
Or you may be imaginary
You may be an integer
You may be finite
And withing limits
Or you may be infinite
Boundless and think that is natural
Because the alternative is not
But you all have to obey some rules

You may be an axiom
Taken for granted
Never needing a justification
Like a King for life
But you have to obey some rules

You may be a conjecture
And think you make sense
But you have no choice
But to depend on someone to prove you right
And you may never be proven right
Forever wandering
In the wind of uncertainty

You may be a theorem
Proven by reasoning
You may be a theory
Thoroughly tested
But you still have to obey some rules

You may be an integral
You may be a derivative
But you both guys need to get it together

You may be
A hidden signal
In a complicated sequence
But you cannot hide
Because hidden or not
You obey some rules

You may be
The square root of a number
But you still
Need to be squared
to be that number

You may be a proposition
And argue that 5 is greater than 4
You can be either true or false

But you still have to make sense

You may be a function
That relates different variables
That acts like a machine
Taking inputs and producing outputs
Which means that you definitely
Obey some rules

You may be linear
And be predictable
Or you may be nonlinear
And be unpredictable
But no matter what
You both obey some rules

You may think you are π
And have both
Randomness and infinity in your genes
But just look around you
You are bound to a circle
The most perfect object in geometry
So, what are your rules?

And even if
You are completely random
And think that
Rules don't apply to you
You still are
Part of the Universe
Thus, there must be
A reason for you

o r n gd

m n tm
e
e

uarrcmtb
d
e
s

ro
oso l
y
lth
eeaa
ere
s
sa s i o

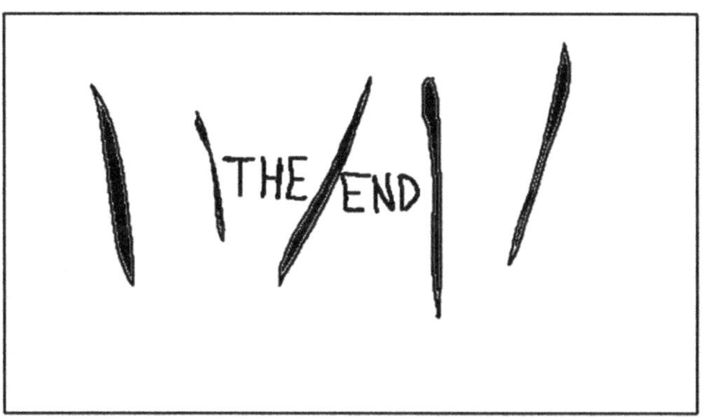

About the author

ANASTASIOS TSONIS is an Emeritus Distinguished Professor in the Department of Mathematical Sciences at the University of Wisconsin-Milwaukee. He studied Mathematics/Physics at the Aristotle University of Thessaloniki, and then he obtained his Doctoral degree in Atmospheric Sciences from McGill University in Montreal, Canada. Subsequently, he was awarded, a post-doctoral fellowship with Atmospheric Environment Service of the government of Canada. After that he became a professor at the University of Wisconsin-Milwaukee, where he built an internationally acclaimed program of atmospheric sciences. His research interests include Chaos theory, Climate dynamics, and Networks. He is the author of more than 140 peer reviewed scientific publications and of eleven books. Four of these books are research books, two are textbooks (one translated into Chinese), two are popular science books (entitled *Randomnicity: Rules and Randomness in the Realm of the*

Infinite and *Aristotle's Μετεωρολογικά: Weather then and now*), one is a translation into English of the epic Greek poem *Kyra Frossini*, and one is a novel entitled *Parallel.* The eleventh is this blend of poems, personal experiences, science, math, photography, and art. He is the winner of the American Geophysical Union's prestigious Ed Lorenz lecture award for significant contributions to nonlinear geoscience. Recently he was elected corresponding member of the Academy of Athens.

Image Credits

Cover: Kihei Beach, Maui. Photo by Anastasios Tsonis

The soul carriers: *Agony and Hope*, 2005, by Anastasios Tsonis

Ode to an identical twin: a) photo property of Anastasios Tsonis b) pinterest.com/cesaremoceo

Time: *Space and Time*, 2022, by Anastasios Tsonis
The game of the name: *Just Names*, 2023, by Anastasios Tsonis

Dream: *Christina's World*, 1948, by Andrew Wyeth, Digital Image © Museum of Modern Art/Licensed by SCALA/Art Resource, NY.

Abstract song: *Birth of a Woman,* 1990, by E. Jungowska. Collection of Anastasios Tsonis

Scars: Photo by Anastasios Tsonis

Random thoughts: *High School Desk,* 1970, lithograph edition of 57, by Willem de Kooning. Previously owned by Anastasios Tsonis

Dream: Illustration from the book *"A Monster Calls"*. Courtesy of Jim Kay

Gallery room: *Untitled*, 1985, woodcut edition of 7, by Per Kirkeby. Collection of Anastasios Tsonis

Time: a) Orbit of Earth around the Sun. Sketch by Anastasios Tsonis. b) Orbit in a three-body problem. Courtesy of Professor Heinz-Otto Peitgen. c) *Past, Present, and Future,* 2013, by Anastasios Tsonis

In their nature: From the German cloud atlas *"Wolken und andere Erscheinungen…"*, by Thomas Forster, 1819, figure 5.

Information on trial: a) Anastasios Tsonis. b) NASA.

Random thoughts: *Untitled*, 1967, etching edition of 144, by Lee Bontecou. Collection of Anastasios Tsonis

Hahaha: *Girl in Sunlight*, 1996, by Dan Muller Collection of Anastasios Tsonis

Random song: clipdealer.com

Imagine: first published in Thomas Wright in "*An Original Theory of the Universe*", 1750

Plus, or minus?: *Vibeke*, 1983, by Per Kirkeby, from Per Kirkeby Painting, Poul Erik Tojner, Editor, 2003

Ode to an identical twin: Photo by Anastasios Tsonis

Two random lives: Writings by Panagiotis and Anastasios Tsonis

Dream: *Dream,* 2022, by Anastasios Tsonis

Untitled: *Flower*, 2005, by Anastasios Tsonis

Nighthawks: *Nighthawks,* 1942, by Edward Hopper. The Art Institute of Chicago/Art Resource, NY.

Untitled: From a mural in Erice, Sicily, 2019. Photo by Anastasios Tsonis

Entanglement: yazzoom.com

Clouds (or, an alternative to weather reporting): Photo by Anastasios Tsonis

The leaf: 2023, by BBB

Ode to Bob Dylan: borealcm.com

The end: *Scars, 2023,* by Anastasios Tsonis (inspired by the artwork of Lucio Fontana)

Or: *Sculpture Park*, 2000, etching, by Franz Ackermann. Collection of Anastasios Tsonis

For the full Dos Madres Press catalog:
www.dosmadres.com

www.ingramcontent.com/pod-product-compliance
Lightning Source LLC
LaVergne TN
LVHW050922210926
848482LV00012B/507
* 9 7 8 1 9 5 3 2 5 2 9 6 8 *